Hydrology and Water Resources: A Comprehensive Questions and Answers Guide

C. P. Kumar

Former Scientist 'G'
National Institute of Hydrology
Roorkee - 247667, India

April 2023

CONTENTS

PREFACE

Water is a precious resource that sustains life on Earth. Hydrology and water resources engineering are essential fields of study that help us understand and manage this vital resource. This book aims to provide a comprehensive collection of questions and answers related to hydrology, water resources, and related topics. The book covers a wide range of topics, including surface water, groundwater, water quality, water resources management, remote sensing and GIS applications in hydrology and water resources, and the impact of climate change on water resources.

This book is intended to be a useful resource for students, researchers, and professionals working in the field of hydrology and water resources. The book is organized into chapters, with each chapter covering a specific topic. Each chapter contains a set of questions and answers, to help readers understand the concepts. The aim is to provide readers with a comprehensive understanding of the subject, from the basics to the latest developments. A chapter has been exclusively devoted for water resources of India.

In addition, this book is also an excellent resource for individuals preparing for written tests and interviews in the field of hydrology and water resources. The questions and answers provided in the book cover a broad spectrum of topics, allowing readers to enhance their knowledge and improve their performance in such assessments. With its comprehensive coverage, the book is an invaluable tool for those seeking to gain a competitive edge in the job market or enhance their career prospects. The book can serve as a self-study guide or as a reference for those working in the field. Overall, this book is a must-have for anyone

interested in hydrology and water resources, whether for
academic, professional, or personal reasons.

C. P. Kumar

Former Scientist 'G'
National Institute of Hydrology
Roorkee - 247667, India

E-mail: cpkumar@yahoo.com
Web: https://www.angelfire.com/nh/cpkumar/

1. INTRODUCTION

Water is one of the most essential resources for the sustenance of life on earth. Hydrology and water resources are interdisciplinary fields that involve the study of the occurrence, distribution, movement, and quality of water in the earth's atmosphere, on the land surface, and underground. The availability of water resources is critical for the survival of both humans and ecosystems. Therefore, it is essential to have a good understanding of the principles and practices of hydrology and water resources management.

This book aims to provide a comprehensive overview of hydrology and water resources, with a particular focus on the water resources of India. The book covers a wide range of topics, including surface water, groundwater, surface water modelling, groundwater modelling, water quality modelling, water resources management, application of remote sensing and GIS in hydrology and water resources, and the impact of climate change on water resources.

The book is structured in a question-and-answer format, making it easy for readers to navigate and find the information they need. Each chapter provides a brief introduction to the topic, followed by a series of questions and answers that cover the key concepts, methods, and applications of the topic. The questions and answers are designed to be accessible to readers with different levels of background knowledge, from beginners to advanced researchers and practitioners.

The book is intended for students, researchers, and practitioners in hydrology, water resources, environmental science, civil engineering, and related fields. It is also suitable for anyone interested in learning about the principles and practices of hydrology and water resources management. I hope that this book will serve as a useful resource for readers interested in understanding the complexities of water resources and the challenges of managing them sustainably.

2. HYDROLOGY

Hydrology is the scientific study of water in the Earth's system, including its distribution, circulation, and properties. It involves the exploration of the occurrence, movement, and quality of water in the atmosphere, surface, and underground. Hydrology is a multidisciplinary field that includes elements of geology, physics, chemistry, biology, and mathematics. Understanding the principles of hydrology is essential for a wide range of applications, including water resources management, flood control, environmental protection, agriculture, and engineering design.

What is hydrology?

Hydrology is the study of water in the earth's system, including its distribution, movement, and quality.

What is the hydrologic cycle?

The hydrologic cycle is the process by which water moves between the earth's surface and the atmosphere.

What are the components of the hydrologic cycle?

The components of the hydrologic cycle include precipitation, evaporation, transpiration, infiltration, runoff, and groundwater flow.

What is precipitation?

Precipitation is any form of water that falls from the atmosphere and reaches the earth's surface, such as rain, snow, sleet, or hail.

What is evaporation?

Evaporation is the process by which water changes from a liquid to a gas and enters the atmosphere.

What is transpiration?

Transpiration is the process by which water is taken up by plants and released into the atmosphere as water vapor.

What is infiltration?

Infiltration is the process by which water seeps into the ground and becomes groundwater.

What is runoff?

Runoff is the flow of water over the surface of the earth, usually in streams, rivers, or lakes.

What is groundwater?

Groundwater is water that is stored underground in soil and rock formations.

What is a watershed?

A watershed is an area of land where all the water that falls within it flows to a common point, such as a river or lake.

What is a hydrograph?

A hydrograph is a graph that shows the discharge of a stream or river over time.

What is a flood?

A flood is an overflow of water onto normally dry land, caused by heavy rainfall, snowmelt, or other factors.

What is a drought?

A drought is a prolonged period of below-average precipitation, resulting in a shortage of water.

What is a water balance?

A water balance is an accounting of the inputs and outputs of water in a particular area, including precipitation, evapotranspiration, and runoff.

What is a water budget?

A water budget is a detailed calculation of the inputs, outputs, and storage of water in a given area.

What is a streamflow?

Streamflow is the flow of water in a stream or river, measured in cubic feet per second or cubic meters per second.

What is a stream gauge?

A stream gauge is a device used to measure the flow of water in a stream or river.

What is a floodplain?

A floodplain is the low-lying area adjacent to a river or other water body that is subject to flooding during periods of high water flow.

What is a hydrologic model?

A hydrologic model is a mathematical representation of the hydrologic cycle, used to simulate the movement of water in a given area.

What is a water quality?

Water quality refers to the chemical, physical, and biological characteristics of water, including its suitability for specific uses.

What is the Clean Water Act?

The Clean Water Act is a federal law in the United States that regulates the discharge of pollutants into the nation's surface waters.

What is the Safe Drinking Water Act?

The Safe Drinking Water Act is a federal law in the United States that regulates the quality of public drinking water supplies.

What is a groundwater recharge?

Groundwater recharge is the process by which water seeps into the ground and replenishes the groundwater supply.

What is an aquifer?

An aquifer is a geologic formation that contains water and can yield usable quantities of water to a well or spring.

What is an artesian well?

An artesian well is a well that taps into a confined aquifer, causing the water to flow to the surface under its own pressure.

What is water scarcity?

Water scarcity is a condition in which there is insufficient water to meet the needs of a population or ecosystem.

What is water pollution?

Water pollution is the contamination of water bodies, such as rivers, lakes, or groundwater, with substances that degrade the water quality and make it harmful to humans or aquatic life.

What is eutrophication?

Eutrophication is the process by which a body of water becomes overly enriched with nutrients, leading to excessive plant growth and the depletion of oxygen in the water.

A stormwater management plan is a plan designed to manage the runoff of rainwater from urban areas, with the goal of reducing flooding and improving water quality.

A flood warning system is a system of sensors and monitoring devices used to detect and predict floods, providing advance notice to communities and allowing for appropriate response.

A drought plan is a plan designed to manage the effects of drought, including conservation measures, emergency response plans, and long-term water management strategies.

A water reuse plan is a plan designed to recycle and reuse wastewater for non-potable uses, such as irrigation or industrial processes.

A water conservation plan is a plan designed to reduce water use through measures such as water-efficient appliances, landscaping practices, and education programs.

A water footprint is a measure of the amount of water used to produce goods and services, including both direct and indirect water use.

What is a virtual water?

Virtual water is the water used to produce goods and services, including the water used in the production process and the water used to grow the raw materials.

What is water management?

Water management is the process of planning, developing, and managing water resources to meet the needs of people and ecosystems.

What is water governance?

Water governance is the process of making decisions about water resources, including the allocation of water rights, the management of water quality, and the coordination of water-related policies and programs.

What is integrated water resources management?

Integrated water resources management is an approach to water management that considers the social, economic, and environmental aspects of water resources, and aims to balance competing demands for water.

What is a water user association?

A water user association is a group of water users who collaborate to manage and allocate water resources in a particular area.

What is a river basin organization?

A river basin organization is a government agency or multi-stakeholder group responsible for managing the water resources of a particular river basin.

What is a water conflict?

A water conflict is a dispute over the use or control of water resources, often between different users or stakeholders.

What is a water transfer?

A water transfer is the movement of water from one location to another, usually to meet the water needs of a particular user or region.

What is a desalination plant?

A desalination plant is a facility that removes salt and other minerals from seawater or brackish water to produce freshwater.

What is a water treatment plant?

A water treatment plant is a facility that treats and purifies water to make it safe for drinking or other uses.

What is a water storage facility?

A water storage facility is a structure or system used to store water, such as a reservoir, tank, or aquifer.

What is a riparian zone?

A riparian zone is the area along the banks of a river or other water body, characterized by unique plant and animal communities.

3. WATER RESOURCES

Water resources refer to the natural water sources that are available for use, including rivers, lakes, groundwater, and oceans. These resources are essential for human survival and are used for a variety of purposes, such as drinking, agriculture, industry, and energy production. Managing water resources is a critical task, particularly in areas where water is scarce or where there is a high demand for water. Issues related to water resources management include water quality, water allocation, water conservation, and water pollution. As the world's population continues to grow and climate change alters precipitation patterns, the sustainable management of water resources is becoming increasingly important.

What is water resources management?

Water resources management is the process of planning, developing, and managing the use of water resources for various purposes, such as drinking water, irrigation, and industrial processes.

What is the importance of water resources?

Water resources are essential for sustaining life and supporting human activities, such as agriculture, industry, and transportation. They also provide habitat for aquatic plants and animals.

What are the major sources of water?

The major sources of water are surface water, groundwater, and precipitation.

What is surface water?

Surface water is water that is found on the surface of the earth, such as rivers, lakes, and wetlands.

Groundwater is water that is found underground in porous rock formations and aquifers.

Precipitation is water that falls from the atmosphere in the form of rain, snow, sleet, or hail.

The water cycle is the continuous movement of water from the atmosphere to the earth's surface and back to the atmosphere.

Water pollution occurs when harmful substances, such as chemicals and microorganisms, are introduced into water bodies.

Water pollution can have a range of harmful effects, including health problems for humans and animals, damage to aquatic ecosystems, and economic losses.

Wastewater is water that has been used for various purposes, such as flushing toilets, washing clothes, and showering.

Water scarcity occurs when the demand for water exceeds the available supply.

The causes of water scarcity include population growth, climate change, and overuse of water resources.

Water conservation is the practice of using water resources efficiently and reducing water waste.

Water conservation methods include fixing leaks, using water-efficient appliances, and practicing water-wise landscaping.

Water reuse is the practice of using treated wastewater for non-potable purposes, such as irrigation and industrial processes.

Desalination is the process of removing salt and other minerals from seawater to produce freshwater.

Desalination can provide a reliable source of freshwater in areas with limited water resources.

Desalination can be expensive, energy-intensive, and can have negative environmental impacts.

Water quality refers to the chemical, physical, and biological characteristics of water.

Factors that can affect water quality include pollution, natural processes, and human activities.

What is the Clean Water Act?

The Clean Water Act is a federal law in the United States that regulates the discharge of pollutants into water bodies and sets water quality standards.

What is the Safe Drinking Water Act?

The Safe Drinking Water Act is a federal law in the United States that regulates the quality of drinking water.

What is a watershed?

A watershed is an area of land where all the water that falls within it drains to a common point, such as a river or lake.

What is water erosion?

Water erosion is the process by which water removes soil and rock from the earth's surface.

What is a water footprint?

A water footprint is the amount of water used to produce a product or service.

What is virtual water?

Virtual water is the amount of water used in the production of a product that is not directly visible to consumers.

What is water pricing?

Water pricing is the practice of assigning a monetary value to water resources in order to encourage efficient use and conservation.

What is water allocation?

Water allocation is the process of assigning water resources to various users or uses, such as agriculture, industry, and domestic consumption.

What is a water right?

A water right is a legal entitlement to use a certain amount of water from a specific source for a specific purpose.

What is water governance?

Water governance refers to the institutions, policies, and practices that govern water resources and their management.

What is transboundary water management?

Transboundary water management is the management of water resources that cross national boundaries.

What is the role of international law in water resources management?

International law plays a role in managing transboundary water resources and addressing water-related conflicts.

What is a water conflict?

A water conflict is a disagreement over the allocation, use, or management of water resources.

What is water security?

Water security refers to the availability of sufficient and safe water resources to meet the needs of a population.

What is a drought?

A drought is a prolonged period of below-average precipitation that can lead to water scarcity and other negative impacts.

What is a flood?

A flood is an overflow of water that can cause damage to infrastructure, property, and human life.

What is a stormwater management system?

A stormwater management system is a system that manages the flow of rainwater and other precipitation to prevent flooding and protect water quality.

What is water treatment?

Water treatment is the process of removing contaminants from water to make it safe for human consumption.

What is a water utility?

A water utility is a public or private entity that provides water services, such as treatment, distribution, and billing.

What is a water distribution system?

A water distribution system is a network of pipes, valves, and other infrastructure that delivers water from a treatment plant to consumers.

What is a water supply system?

A water supply system is a system that provides water resources to a community or region.

What is water productivity?

Water productivity refers to the amount of economic output generated per unit of water used.

What is integrated water resources management?

Integrated water resources management is a holistic approach to managing water resources that considers the social, economic, and environmental aspects of water use.

What is a water balance?

A water balance is an accounting of the inflows and outflows of water in a particular area.

What is a water audit?

A water audit is a detailed analysis of water use and losses in a particular system, such as a building or a community.

What is a water management plan?

A water management plan is a comprehensive plan for managing water resources in a particular area.

What is a water conservation plan?

A water conservation plan is a plan for reducing water use and promoting water conservation in a particular area.

What is a water efficiency program?

A water efficiency program is a program that promotes the use of water-efficient practices and technologies to reduce water use.

What is a water footprint assessment?

A water footprint assessment is a method for measuring the amount of water used to produce a product or service.

What are some of the major water resource problems that we face today?

Water scarcity, water pollution, and climate change-induced droughts and floods are some of the major water resource problems we face.

How can we address water scarcity?

We can address water scarcity by adopting water conservation measures, promoting rainwater harvesting, and investing in new water supply infrastructure.

What are the main sources of water pollution?

The main sources of water pollution are industrial discharge, agricultural runoff, sewage and wastewater discharge, and plastic waste.

Measures to prevent water pollution include strict industrial regulations, sustainable agricultural practices, improved sewage treatment, and reducing plastic waste.

We can improve water quality in rivers and lakes by implementing watershed management practices, reducing nutrient pollution, and restoring wetlands and riparian habitats.

We can address climate change-induced droughts and floods by developing water storage and management infrastructure, promoting water-efficient agricultural practices, and reducing greenhouse gas emissions.

Water conservation is the practice of using water efficiently and reducing waste. It is important because it helps to preserve scarce water resources and reduce the need for new water supply infrastructure.

We can encourage people to conserve water through public education campaigns, financial incentives for water-efficient practices, and regulations that require water conservation measures.

Technology can play a key role in solving water resource problems by developing new water treatment and desalination technologies, improving irrigation efficiency, and developing smart water management systems.

What are some of the benefits of sustainable water management practices?

Benefits of sustainable water management practices include increased water security, improved water quality, reduced greenhouse gas emissions, and economic benefits from water-efficient practices.

What is water education?

Water education is the practice of educating the public about water resources and promoting water conservation and sustainability.

4. WATER RESOURCES OF INDIA

India is a country with a rich and diverse water resource base. It is home to major rivers like the Ganges, Brahmaputra, Yamuna, Godavari, Krishna, and Cauvery, which form the backbone of the country's agricultural and industrial sectors. Additionally, India has an extensive network of lakes, ponds, and groundwater reserves that provide crucial water resources to millions of people. However, the management of these water resources has been a significant challenge for the country due to a range of factors such as population growth, urbanization, climate change, and inadequate infrastructure. As a result, the Indian government and other stakeholders have been actively working to address these challenges and ensure sustainable use and management of water resources for the country's socio-economic development.

What percentage of India's land area is covered by water bodies?
Approximately 4% of India's land area is covered by water bodies.

What are the major rivers in India?
The major rivers in India are the Ganges, Brahmaputra, Indus, Yamuna, Godavari, Krishna, Narmada, and Mahanadi.

What is the length of the Ganges river in India?
The Ganges river in India is approximately 2,525 kilometers long.

Which river is known as the "Sorrow of Bengal"?
The Brahmaputra river is known as the "Sorrow of Bengal" due to its tendency to flood.

What is the total renewable water resources of India?

The total renewable water resources of India are estimated to be 1,897.4 billion cubic meters (BCM) per year.

What percentage of India's population has access to safe drinking water?

As of 2021, approximately 89% of India's population has access to safe drinking water.

What is the main source of water for agriculture in India?

Groundwater is the main source of water for agriculture in India.

What is the largest dam in India?

The Sardar Sarovar Dam on the Narmada River is currently the largest dam in India.

Which state in India has the highest percentage of irrigated land?

Punjab has the highest percentage of irrigated land in India, with over 98% of its cropped area being irrigated.

What is the name of India's national river?

The Ganges river is considered India's national river.

What is the main cause of water pollution in India?

Industrial and agricultural activities are the main causes of water pollution in India.

What is the percentage of groundwater depletion in India?

As of 2021, the rate of groundwater depletion in India is estimated to be around 61%.

Wular Lake in Jammu and Kashmir is the largest freshwater lake in India.

Maharashtra has the highest number of dams in India, with over 3,000 dams.

Approximately 12% of India's land area is prone to floods.

As of 2021, the estimated per capita water availability in India is around 1,545 cubic meters per year.

India's first floating solar power plant is located at the Banasura Sagar reservoir in Kerala.

The main cause of water scarcity in India is the uneven distribution of water resources and mismanagement of available water resources.

Uttarakhand has the highest number of hot springs in India.

Govind Ballabh Pant Sagar, also known as Rihand Dam Reservoir, is the largest artificial lake in India.

Majuli Island on the Brahmaputra river is the largest river island in India.

What is the percentage of India's population that depends on agriculture for their livelihood?

Approximately 58% of India's population depends on agriculture for their livelihood

What is the name of the project aimed at linking India's major rivers?

The project aimed at linking India's major rivers is called the Interlinking of Rivers (ILR) project.

What is the percentage of India's surface water that is utilized for irrigation?

Approximately 84% of India's surface water is utilized for irrigation.

What is the name of the largest man-made lake in India?

The Hirakud Dam Reservoir in Odisha is the largest man-made lake in India.

Which state in India has the highest number of wetlands?

West Bengal has the highest number of wetlands in India.

What is the percentage of India's population that practices open defecation?

As of 2021, the percentage of India's population that practices open defecation is estimated to be around 30%.

What is the name of the largest waterfall in India?

Jog Falls in Karnataka is the largest waterfall in India.

What is the percentage of India's population that is dependent on groundwater for drinking water?

Approximately 85% of India's rural population and 48% of its urban population is dependent on groundwater for drinking water.

The Central Water Commission (CWC) is the organization responsible for managing water resources in India.

Approximately 38% of India's water resources are shared with neighboring countries.

The Indus river forms the border between India and Pakistan.

Meghalaya has the highest number of waterfalls in India.

The Polavaram Dam is the largest dam on the Godavari river.

Approximately 60% of India's groundwater is used for irrigation.

Chilika Lake in Odisha is the largest saltwater lake in India.

The Ganges river flows through the Sundarbans delta.

What is the name of the largest waterfall in India by volume of water?

The Nohkalikai Falls in Meghalaya is the largest waterfall in India by volume of water.

What is the percentage of India's surface water that is utilized for domestic purposes?

Approximately 8% of India's surface water is utilized for domestic purposes.

What is the name of the river that flows through Delhi?

The Yamuna river flows through Delhi.

Which state in India has the highest number of glaciers?

Jammu and Kashmir has the highest number of glaciers in India.

What is the name of the largest dam on the Krishna river?

The Nagarjuna Sagar Dam is the largest dam on the Krishna river.

What is the percentage of India's population that lives in areas with high water stress?

As of 2021, approximately 54% of India's population lives in areas with high water stress.

What is the name of the river that flows through Mumbai?

The Mithi river flows through Mumbai.

Which state in India has the highest number of natural lakes?

Jammu and Kashmir has the highest number of natural lakes in India.

What is the name of the river that flows through Varanasi?

The Ganges river flows through Varanasi.

What is the name of the largest dam on the Tungabhadra river?

The Tungabhadra Dam is the largest dam on the Tungabhadra river.

What is the percentage of India's surface water that is utilized for industrial purposes?

Approximately 8% of India's surface water is utilized for industrial purposes.

What is the name of the river that forms the border between India and Bangladesh?

The Brahmaputra river forms the border between India and Bangladesh.

What is the percentage of India's groundwater that is used for domestic purposes?

Approximately 20% of India's groundwater is used for domestic purposes.

What is the major cause of water scarcity in India?

Overexploitation of groundwater resources and poor water management.

How can water scarcity be addressed in India?

By implementing water conservation and management practices, and investing in water infrastructure.

What are the consequences of water scarcity in India?

Decreased agricultural productivity, water-borne diseases, and social conflicts.

How can we improve water quality in India?

By treating wastewater, improving sanitation, and reducing pollution.

What is the impact of climate change on water resources in India?

Increased frequency and intensity of droughts and floods.

How can we increase water use efficiency in agriculture in India?

By promoting efficient irrigation technologies, crop diversification, and rainwater harvesting.

What are the challenges of inter-state river disputes in India?

Political and legal issues, water sharing conflicts, and lack of effective management mechanisms.

How can we reduce water consumption in urban areas in India?

By promoting water conservation practices, reducing leakage and theft, and implementing efficient water pricing policies.

What is the role of community participation in water resource management in India?

Community involvement can help in identifying local water needs, implementing water management plans, and improving water governance.

How can we ensure sustainable management of water resources in India?

By adopting integrated water resource management approaches, promoting conservation and re-use, and investing in water infrastructure.

What is the main problem associated with surface water in India?

The main problem is contamination with pollutants like sewage and industrial waste.

What are the sources of pollution in surface water in India?

Sources of pollution include agricultural runoff, urban and industrial waste, and improper disposal of solid waste.

How does pollution in surface water affect human health in India?

Pollution in surface water can cause waterborne diseases like cholera, typhoid, and hepatitis A.

What are the government initiatives in India to address surface water pollution?

Initiatives include the National River Conservation Plan and the Clean Ganga Mission.

What is the impact of climate change on surface water in India?

Climate change can cause droughts and floods, affecting the quantity and quality of surface water.

How does groundwater depletion affect surface water in India?

Groundwater depletion reduces surface water availability, as surface water is replenished by groundwater.

What are the measures to conserve surface water in India?

Measures include rainwater harvesting, watershed management, and promoting water-efficient agricultural practices.

How does deforestation affect surface water in India?

Deforestation reduces the ability of soil to hold water, leading to soil erosion and decreased surface water availability.

What are the challenges in managing surface water resources in India?

Challenges include competing demands for water, inadequate infrastructure, and weak regulatory enforcement.

How can individuals contribute to the conservation of surface water in India?

Individuals can conserve water by reducing water use, proper waste disposal, and supporting sustainable water management practices.

What is the main cause of groundwater depletion in India?

Over-extraction for agriculture, industry and domestic purposes.

How can groundwater depletion be prevented?

By adopting sustainable groundwater management practices such as rainwater harvesting and water-use efficiency.

What are the effects of groundwater depletion on India's agriculture?

Decreased crop yields, increased input costs, and farmer distress.

What are the health implications of drinking contaminated groundwater in India?

Risk of waterborne diseases like cholera, typhoid, and hepatitis.

What is the primary source of groundwater contamination in India?

Agricultural runoff and industrial waste discharge.

How can groundwater contamination be reduced?

By implementing proper waste disposal mechanisms and treatment of effluents before discharge.

What is the role of government in addressing groundwater-related problems in India?

Implementing policies and regulations to ensure sustainable groundwater use and management.

How can groundwater recharge be improved in India?

By constructing recharge structures, promoting watershed development, and conserving soil moisture.

What are the socio-economic impacts of groundwater depletion in India?

Decreased livelihood opportunities, increased poverty, and rural-urban migration.

What is the importance of groundwater conservation in India?

It is a critical source of drinking water and irrigation for millions of people, and its depletion threatens both socio-economic and ecological sustainability.

What are the major challenges faced in water resources management in India?

Water scarcity, groundwater depletion, pollution, and inefficient use of water resources.

How can India address the issue of water scarcity?

By promoting water conservation, implementing rainwater harvesting, and improving water use efficiency in agriculture.

What are the main sources of water pollution in India?

Industrial and municipal wastewater, agricultural runoff, and solid waste disposal.

How can India tackle water pollution?

By enforcing stricter regulations, increasing public awareness, and implementing effective waste management practices.

What are the consequences of groundwater depletion in India?

Increased reliance on surface water, decreased agricultural productivity, and land subsidence.

How can India conserve groundwater?

By promoting efficient irrigation techniques, recharging aquifers through rainwater harvesting, and implementing groundwater regulation.

How can India improve the efficiency of water use in agriculture?

By promoting drip irrigation and other efficient irrigation technologies, improving soil health, and promoting crop diversification.

What is the role of community participation in water resources management in India?

Community participation can lead to better management of water resources, improved access to water, and increased awareness about water conservation.

By establishing strong institutional frameworks, promoting stakeholder participation, and enforcing accountability mechanisms.

Inter-state cooperation is essential for effective management of shared river basins, equitable allocation of water resources, and resolution of disputes.

5. SURFACE WATER

Surface water refers to any water that is above the surface of the Earth, such as in rivers, lakes, reservoirs, and oceans. It is an essential resource for many human activities, including drinking water, agriculture, industry, and transportation. Surface water is also an important part of the Earth's natural ecosystem, providing habitat for plants and animals. The quality and quantity of surface water can be affected by natural processes such as precipitation and erosion, as well as human activities such as land use changes, pollution, and dam construction. Management and conservation of surface water resources are critical for ensuring sustainable development and protection of the environment.

What is surface water?

Surface water refers to water that is found on the Earth's surface, such as in rivers, lakes, and oceans.

What is the difference between surface water and groundwater?

Surface water is water that is found on the Earth's surface, while groundwater is water that is found underground in aquifers.

What is the importance of surface water?

Surface water is important for human consumption, irrigation, industry, transportation, and recreation.

What are the sources of surface water?

The sources of surface water include precipitation, snowmelt, and groundwater discharge.

What is the water cycle?

The water cycle is the continuous movement of water on, above, and below the Earth's surface. It involves processes such as evaporation, precipitation, and transpiration.

What is the role of surface water in the water cycle?

Surface water plays an important role in the water cycle as it is a major component of precipitation, evaporation, and transpiration.

What is a river?

A river is a large natural stream of water that flows through a channel to the sea, a lake, or another river.

What is a lake?

A lake is a large body of water that is surrounded by land.

What is a reservoir?

A reservoir is an artificial lake that is created by constructing a dam across a river or other waterway.

What is a wetland?

A wetland is an area of land that is saturated with water, such as a marsh or swamp.

What is a watershed?

A watershed is an area of land where all of the water that falls within it drains into a common body of water, such as a river or lake.

What is the importance of watersheds?

Watersheds are important because they provide drinking water, support wildlife and fisheries, and regulate the flow of water.

What is runoff?

Runoff is the water that flows over the land surface during and after precipitation.

What is erosion?

Erosion is the process of wearing away or removing soil and rock from the Earth's surface.

What is sedimentation?

Sedimentation is the process of settling or deposition of particles in water.

What is a floodplain?

A floodplain is an area of land adjacent to a river or stream that is subject to flooding.

What is a delta?

A delta is a landform that is created by the deposition of sediment at the mouth of a river.

What is a levee?

A levee is a man-made embankment that is built along a river to prevent flooding.

What is the flood frequency?

Flood frequency is the probability of a flood occurring in a given area over a specified time period.

What is the floodplain management?

Floodplain management is the process of using land-use planning and zoning to reduce the risk of flooding in flood-prone areas.

What is a dam?

A dam is a structure that is built across a river or other waterway to regulate the flow of water.

What is the purpose of a dam?

The purpose of a dam is to store water, generate electricity, and control flooding.

What is a spillway?

A spillway is a structure that is built into a dam to allow excess water to flow out.

What is a fish ladder?

A fish ladder is a series of pools or steps that are built next to a dam or other obstruction in a river to allow fish to swim upstream.

What is water quality?

Water quality refers to the physical, chemical, and biological characteristics of water.

What are the parameters used to measure water quality?

Parameters used to measure water quality include pH, temperature, dissolved oxygen, turbidity, and nutrients.

What is the Clean Water Act?

The Clean Water Act is a federal law in the United States that regulates the discharge of pollutants into surface waters.

What is eutrophication?

Eutrophication is the process by which excess nutrients, such as nitrogen and phosphorus, cause excessive growth of algae and other aquatic plants, leading to oxygen depletion and fish kills.

What is thermal pollution?

Thermal pollution is the discharge of heated water into a body of water, which can cause changes in water temperature and harm aquatic life.

What is acid rain?

Acid rain is a type of precipitation that contains acidic compounds, such as sulfuric acid and nitric acid, which can harm aquatic and terrestrial ecosystems.

What is point source pollution?

Point source pollution is pollution that comes from a single identifiable source, such as a factory or sewage treatment plant.

What is nonpoint source pollution?

Nonpoint source pollution is pollution that comes from diffuse sources, such as agricultural runoff or urban stormwater runoff.

What is a Total Maximum Daily Load (TMDL)?

A Total Maximum Daily Load (TMDL) is a calculation of the maximum amount of a pollutant that a water body can receive and still meet water quality standards.

What is the role of wetlands in water quality?

Wetlands are important in water quality because they act as natural filters, removing pollutants and excess nutrients from water.

What is the impact of climate change on surface water?

Climate change is expected to have significant impacts on surface water, including changes in precipitation patterns, increased frequency and intensity of floods and droughts, and changes in water quality.

A lake is a natural body of water, while a reservoir is an artificial body of water created by building a dam.

A river is a large natural waterway that flows through a channel, while a stream is a smaller natural waterway that flows into a larger waterway.

A flood is a slow-rise and prolonged inundation of an area, while a flash flood is a sudden and rapid rise in water levels due to heavy rainfall.

An estuary is a body of water where a river meets the sea and freshwater mixes with saltwater, while a delta is a landform created by the deposition of sediment at the mouth of a river.

A weir is a low structure built across a river to measure or regulate flow, while a dam is a high structure built to store water or generate electricity.

Water quantity refers to the amount of water available, while water quality refers to the physical, chemical, and biological characteristics of the water.

Wetlands are areas of land where water covers the soil, while swamps are wetlands dominated by trees and woody vegetation.

What is the difference between a canal and a channel?

A canal is an artificial waterway built for navigation or irrigation, while a channel is a natural or artificial waterway that directs the flow of water.

What is the difference between a riffle and a pool in a stream?

A riffle is a shallow area with faster-moving water and a rocky bottom, while a pool is a deeper area with slower-moving water and a smooth bottom.

What is the difference between a perennial stream and an intermittent stream?

A perennial stream flows continuously throughout the year, while an intermittent stream flows only during certain times of the year or after rainfall.

What is the difference between a lagoon and a bay?

A lagoon is a shallow body of water separated from a larger body of water by a barrier, while a bay is a broad inlet of the sea where the land curves inward.

What is the difference between a reservoir and a cistern?

A reservoir is a large artificial body of water created by building a dam, while a cistern is a small tank used for storing rainwater or other water.

What is the difference between a floodplain and a levee?

A floodplain is a flat area near a river or other body of water that is subject to flooding, while a levee is a raised embankment built along a river to prevent flooding.

What is the difference between a spring and a seep?

A spring is a natural outflow of water from the ground, while a seep is a slow and continuous flow of water through the ground.

6. GROUNDWATER

Groundwater refers to the water that is found beneath the surface of the Earth, in the spaces between rocks and soil particles in underground aquifers. It is a vital resource for many human activities, such as irrigation, drinking water supply, and industrial uses. Groundwater is replenished through natural processes such as rainfall and snowmelt, and it moves slowly through the subsurface, often taking years to travel from its source to a discharge point such as a spring, river or well. The quality and quantity of groundwater can be affected by natural processes such as geology, climate, and land use, as well as human activities such as contamination and over-extraction. Management and protection of groundwater resources are crucial for ensuring sustainable water supply and preventing environmental degradation.

What is groundwater?
Groundwater is water that is found beneath the earth's surface in the pore spaces of soil, sand, and rock formations.

What is an aquifer?
An aquifer is an underground layer of permeable rock, sediment, or soil that contains water.

How does groundwater form?
Groundwater forms when rain or snowmelt infiltrates into the ground and seeps downward until it reaches a point where the soil and rock are saturated with water.

What are the benefits of groundwater?
Groundwater is an important source of drinking water, irrigation, and industrial uses.

What is a water table?

The water table is the upper surface of the zone of saturation, where water is present in the soil or rock.

What is an artesian well?

An artesian well is a well in which water flows naturally to the surface under pressure from an underground aquifer.

How is groundwater pumped to the surface?

Groundwater is pumped to the surface using a well, which is a hole drilled into the ground to access the underground aquifer.

What is the recharge zone of an aquifer?

The recharge zone of an aquifer is the area where water infiltrates into the ground and replenishes the groundwater.

What is a groundwater basin?

A groundwater basin is a geologic formation that contains one or more interconnected aquifers.

What is a confined aquifer?

A confined aquifer is an aquifer that is bounded above and below by impermeable rock or sediment, which restricts the flow of water.

What is a perched aquifer?

A perched aquifer is a small, localized aquifer that is separated from a larger, regional aquifer by an impermeable layer of rock or sediment.

What is a wellhead protection area?

A wellhead protection area is the area around a well that is designated for protection to prevent contamination of the groundwater supply.

What is groundwater contamination?

Groundwater contamination occurs when pollutants from human activities enter the groundwater and make it unsafe to drink or use.

What are some sources of groundwater contamination?

Some sources of groundwater contamination include leaking underground storage tanks, landfills, industrial activities, and agricultural practices.

What is the Clean Water Act?

The Clean Water Act is a federal law in the United States that regulates the discharge of pollutants into the nation's surface waters and sets water quality standards.

What is the Safe Drinking Water Act?

The Safe Drinking Water Act is a federal law in the United States that regulates the quality of public drinking water supplies and sets standards for contaminants.

What is a contaminant plume?

A contaminant plume is a cloud of contaminated groundwater that moves through the subsurface.

What is remediation?

Remediation is the process of cleaning up contaminated groundwater to make it safe to drink or use.

What are some remediation techniques?

Some remediation techniques include pump and treat, air sparging, and in situ bioremediation.

What is pump and treat?

Pump and treat is a remediation technique in which groundwater is pumped to the surface, treated to remove contaminants, and then re-injected into the ground.

What is air sparging?

Air sparging is a remediation technique in which air is injected into the ground to enhance the biodegradation of contaminants.

What is in situ bioremediation?

In situ bioremediation is a remediation technique in which microorganisms are used to degrade contaminants in the subsurface.

What is a groundwater monitoring network?

A groundwater monitoring network is a system of wells and sampling points that are used to monitor the quality and quantity of groundwater over time.

What is groundwater modelling?

Groundwater modelling is the process of using computer models to simulate the movement of groundwater through the subsurface and predict future conditions.

What is the importance of groundwater modelling?

Groundwater modelling is important for understanding how groundwater resources are affected by human activities, climate change, and other factors.

What is the Ogallala Aquifer?

The Ogallala Aquifer is a large, shallow aquifer located beneath the Great Plains of the United States, which is an important source of irrigation water for agriculture.

Groundwater depletion occurs when the rate of groundwater pumping exceeds the rate of recharge, leading to a decline in the water table.

The causes of groundwater depletion include over-pumping for irrigation, urbanization, and climate change.

The consequences of groundwater depletion include land subsidence, reduced water availability for drinking and irrigation, and ecological impacts.

Groundwater recharge is the process of replenishing the groundwater through the infiltration of rain or snowmelt.

The methods of groundwater recharge include infiltration basins, injection wells, and spreading grounds.

An infiltration basin is a depression in the ground that is designed to capture and infiltrate stormwater runoff, which helps to recharge the groundwater.

An injection well is a well that is used to inject treated wastewater or other fluids into the subsurface for the purpose of groundwater recharge.

A spreading ground is a site where water is spread over the ground surface to allow for infiltration and recharge of the groundwater.

What is a groundwater management plan?

A groundwater management plan is a comprehensive plan for the sustainable management of a groundwater basin, which includes measures to balance water supply and demand, protect water quality, and ensure long-term sustainability.

What is a water right?

A water right is a legal right to use a certain quantity of water for a specific purpose, such as irrigation or municipal supply.

What is groundwater law?

Groundwater law is the body of law that governs the use, management, and protection of groundwater resources.

What is a groundwater district?

A groundwater district is a special purpose governmental entity that is responsible for managing and protecting groundwater resources within a specific geographic area.

What is a groundwater recharge area?

A groundwater recharge area is an area where water infiltrates into the ground and recharges the groundwater.

What is a groundwater dependent ecosystem?

A groundwater dependent ecosystem is an ecosystem that relies on groundwater as its primary source of water.

What are some examples of groundwater dependent ecosystems?

Some examples of groundwater dependent ecosystems include wetlands, riparian areas, and desert springs.

What is a well interference zone?
A well interference zone is the area around a well where pumping can cause a reduction in the water level or flow of nearby wells.

What is a groundwater sustainability agency?
A groundwater sustainability agency is a new type of governmental entity created under California law to manage and regulate groundwater resources in critically over-drafted basins.

What is a groundwater recharge credit?
A groundwater recharge credit is a credit given to a user who contributes to groundwater recharge, which can be redeemed later to offset pumping costs or to meet regulatory requirements.

What is a wellhead protection area?
A wellhead protection area is the area around a well that is designated for protection to prevent contamination of the groundwater supply.

What is a cone of depression?
A cone of depression is a depression in the water table around a well that results from pumping, which can cause nearby wells to go dry.

What is groundwater banking?
Groundwater banking is the practice of storing water underground during wet periods for use during dry periods, which can help to manage water resources and reduce the impacts of drought.

What is a groundwater contamination plume?
A groundwater contamination plume is an area of contaminated groundwater that extends downgradient from

a source of contamination, such as a landfill or industrial site.

What is groundwater remediation?

Groundwater remediation is the process of treating and cleaning up contaminated groundwater to restore it to a usable condition.

What is the role of groundwater in the hydrologic cycle?

Groundwater is an important component of the hydrologic cycle, as it represents the storage and movement of water through the subsurface. Groundwater also plays a key role in sustaining streams, lakes, and wetlands, and provides a vital source of water for drinking, irrigation, and other uses.

7. SURFACE WATER MODELLING

Surface water modelling is the process of creating mathematical or computational models to simulate the behavior and dynamics of surface water systems, such as rivers, lakes, and watersheds. The models use data such as precipitation, runoff, and water quality to simulate the flow and behavior of surface water, and can be used to make predictions about how surface water systems will respond to changes in the environment or to human activities.

Surface water models can be used for a variety of purposes, such as predicting floods, managing water resources, assessing the impact of land use changes, and designing infrastructure such as dams and levees. These models can help decision-makers to make informed choices about how to manage and protect surface water resources, and can provide valuable information for stakeholders such as water utilities, farmers, and conservation organizations. Surface water modelling is a complex and interdisciplinary field, involving elements of hydrology, hydraulics, geology, ecology, and computer science.

What is surface water modelling?

Surface water modelling is the process of using mathematical and computational models to simulate the behavior of surface water systems, such as rivers, lakes, and wetlands.

Why is surface water modelling important?

Surface water modelling is important because it helps scientists and engineers better understand how surface water systems work and how they are affected by changes in the environment.

What types of models are used in surface water modelling?

There are many types of models used in surface water modelling, including hydrological models, hydraulic models, and water quality models.

What is a hydrological model?

A hydrological model is a type of surface water model that simulates the movement of water through the hydrological cycle, including precipitation, evaporation, infiltration, and runoff.

What is a hydraulic model?

A hydraulic model is a mathematical or physical representation of the behavior of water in a system, such as a river, lake, or watershed. A hydraulic model is used to simulate and analyze the flow of water in the system, including variables such as velocity, depth, and pressure.

What is a water quality model?

A water quality model is a type of surface water model that simulates the transport and fate of pollutants in surface water systems.

What data is needed to develop a surface water model?

The data needed to develop a surface water model includes information about the surface water system being modeled, such as the location, topography, hydrology, and water quality.

How is surface water modelling used in environmental management?

Surface water modelling is used in environmental management to help predict the effects of changes to the environment, such as land use changes or climate change, on surface water systems.

A watershed is an area of land where all the water that falls within it drains to a common point, such as a river or lake.

Surface water modelling is used in watershed management to help identify areas of the watershed that are most susceptible to erosion, flooding, or other water-related issues.

A streamflow gauge is a device used to measure the flow of water in a river or stream.

Streamflow gauges are used in surface water modelling to collect data on the flow of water in rivers and streams, which is used to calibrate and validate surface water models.

A rating curve is a graph that shows the relationship between the height of water in a river or stream and the flow of water.

A rating curve is used in surface water modelling to estimate the flow of water in a river or stream based on measurements of the height of the water.

A floodplain is an area of land that is prone to flooding when a river or stream overflows its banks.

How is surface water modelling used to assess flood risk?

Surface water modelling is used to assess flood risk by simulating the behavior of a river or stream during a flood event and predicting areas that are likely to flood.

What is a stormwater model?

A stormwater model is a type of surface water model that simulates the movement of stormwater runoff through a drainage system, such as a storm sewer or detention pond.

How is a stormwater model used in urban planning?

A stormwater model is used in urban planning to design and size stormwater management systems, such as detention ponds and infiltration basins, to reduce the impacts of urbanization on surface water systems.

What is a sediment transport model?

A sediment transport model is a type of surface water model that simulates the movement of sediment through a river or stream.

How is a sediment transport model used in surface water modelling?

A sediment transport model is used in surface water modelling to predict the erosion and deposition of sediment in rivers and streams, which is important for managing sedimentation and maintaining water quality.

What is a groundwater model?

A groundwater model is a type of model that simulates the movement of water through groundwater systems, such as aquifers.

Groundwater models are related to surface water modelling because groundwater and surface water systems are often interconnected, and changes to one system can affect the other.

A flow duration curve is a graph that shows the percentage of time that a river or stream flows at a certain discharge or flow rate.

A flow duration curve is used in surface water modelling to estimate the frequency and duration of different flow conditions in a river or stream, which is important for managing water resources and predicting flood events.

A hydraulic jump is a phenomenon that occurs when a high-velocity flow of water abruptly slows down and forms a turbulent pool of water.

A hydraulic jump is modeled in surface water modelling using hydraulic models that simulate the flow of water through a river or stream and can predict the occurrence of hydraulic jumps.

A cross-section is a vertical slice through a river or stream that shows the depth of water and the shape of the river bed.

How are cross-sections used in surface water modelling?

Cross-sections are used in surface water modelling to represent the shape of a river or stream in a hydraulic model and to estimate the flow of water at different points along the river or stream.

What is a floodplain map?

A floodplain map is a map that shows the areas that are likely to flood during a major flood event.

How is a floodplain map created?

A floodplain map is created using surface water models that simulate the behavior of a river or stream during a major flood event and predict the areas that are likely to flood.

What is a culvert?

A culvert is a structure that is used to allow water to flow under a road or other obstacle.

How are culverts modeled in surface water modelling?

Culverts are modeled in surface water modelling using hydraulic models that simulate the flow of water through a culvert and can predict the capacity of the culvert and the potential for flooding.

What is a weir?

A weir is a structure that is used to control the flow of water in a river or stream.

How is a weir modeled in surface water modelling?

A weir is modeled in surface water modelling using hydraulic models that simulate the flow of water over the weir and can predict the amount of water flowing through the weir.

What is a levee?

A levee is a raised embankment that is used to protect against flooding.

How are levees modeled in surface water modelling?

Levees are modeled in surface water modelling using hydraulic models that simulate the flow of water during a flood event and can predict the effectiveness of levees in preventing flooding.

What is a dam?

A dam is a barrier that is built across a river or stream to control the flow of water and generate electricity.

How is a dam modeled in surface water modelling?

A dam is modeled in surface water modelling using hydraulic models that simulate the flow of water through the dam and can predict the impact of the dam on the downstream river or stream.

What is a hydrologic cycle?

The hydrologic cycle is the continuous movement of water between the atmosphere, the earth's surface, and underground.

How is the hydrologic cycle related to surface water modelling?

The hydrologic cycle is related to surface water modelling because surface water models simulate the movement of water through rivers, streams, and other surface water systems, which is a key component of the hydrologic cycle.

What is a flood frequency analysis?

A flood frequency analysis is a statistical method used to estimate the likelihood of different flood events based on historical data.

How is a flood frequency analysis used in surface water modelling?

A flood frequency analysis is used in surface water modelling to estimate the frequency and magnitude of different flood events, which is important for managing flood risk and designing flood control systems.

What is a rain gauge?

A rain gauge is a device used to measure the amount of precipitation that falls in a specific area.

How are rain gauges used in surface water modelling?

Rain gauges are used in surface water modelling to collect data on precipitation that can be used to calibrate and validate surface water models and to estimate the amount of water entering surface water systems.

What is a stream gauge?

A stream gauge is a device used to measure the flow rate of water in a river or stream.

How are stream gauges used in surface water modelling?

Stream gauges are used in surface water modelling to collect data on flow rates that can be used to calibrate and validate surface water models and to estimate the amount of water entering and leaving surface water systems.

What is a water quality model?

A water quality model is a type of surface water model that simulates the transport and fate of pollutants in surface water systems.

How is a water quality model used in surface water modelling?

A water quality model is used in surface water modelling to predict the movement of pollutants through surface water systems and to identify strategies for improving water quality.

What is a watershed?

A watershed is an area of land where all of the water that falls within it drains to a common point, such as a river or lake.

How are watersheds used in surface water modelling?

Watersheds are used in surface water modelling to define the boundaries of the surface water system being modeled and to estimate the amount of water entering and leaving the system.

What are some examples of surface water models?

Surface water models include HEC-RAS, SWMM, MIKE 11, and WEAP.

What is HEC-RAS model?

HEC-RAS (Hydrologic Engineering Center's River Analysis System) is a one-dimensional hydraulic model used to simulate water flow in rivers, streams, and channels.

What is SWMM model?

SWMM (Storm Water Management Model) is a one-dimensional urban drainage model used to simulate stormwater runoff, including the quantity and quality of the water.

What is MIKE 11 model?

MIKE 11 is a one-dimensional hydraulic model used to simulate water flow in rivers, streams, and channels.

What is WEAP model?

WEAP (Water Evaluation and Planning System) is a water resources planning model that can simulate the behavior of surface water and groundwater under different scenarios.

8. GROUNDWATER MODELLING

Groundwater modelling is the process of creating mathematical or computational models to simulate the behavior and dynamics of groundwater systems, such as aquifers and underground reservoirs. The models use data such as groundwater levels, recharge rates, and hydraulic conductivity to simulate the flow and behavior of groundwater, and can be used to make predictions about how groundwater systems will respond to changes in the environment or to human activities.

Groundwater models can be used for a variety of purposes, such as predicting the impact of pumping on groundwater resources, assessing the impact of land use changes on groundwater quality, designing remediation strategies for contaminated groundwater, and predicting the movement of contaminants in groundwater systems. These models can help decision-makers to make informed choices about how to manage and protect groundwater resources, and can provide valuable information for stakeholders such as water utilities, regulators, and conservation organizations.

Groundwater modelling is a complex and interdisciplinary field, involving elements of hydrology, geology, chemistry, and computer science. The models must take into account factors such as the geology and topography of the area, the properties of the aquifer, and the sources and types of contaminants that may be present. Groundwater modelling can be a powerful tool for managing and protecting groundwater resources, but it requires careful calibration and validation to ensure that the models accurately reflect the behavior of the groundwater system.

What is groundwater modelling?

Groundwater modelling is the process of creating a mathematical representation of the physical system that governs the flow of groundwater in an aquifer.

Why is groundwater modelling important?

Groundwater modelling helps us understand and manage groundwater resources, predict the impacts of human activities on groundwater systems, and evaluate potential solutions for groundwater-related problems.

What are the types of groundwater models?

The types of groundwater models include analytical models, numerical models, and hybrid models.

What is an analytical groundwater model?

An analytical groundwater model is a model that uses mathematical equations to describe the flow of groundwater in an aquifer.

What is a numerical groundwater model?

A numerical groundwater model is a model that divides the aquifer into a grid of cells and solves equations for each cell to simulate the flow of groundwater.

What is a hybrid groundwater model?

A hybrid groundwater model combines elements of both analytical and numerical models.

What are the inputs to a groundwater model?

The inputs to a groundwater model include geological data, hydrological data, meteorological data, and data on human activities that may affect groundwater resources.

A conceptual model is a simplified representation of the physical system that governs the flow of groundwater in an aquifer.

A numerical grid is a grid of cells that is used to divide the aquifer in a numerical groundwater model.

A boundary condition is a condition that is applied to the boundary of the aquifer in a groundwater model.

A recharge rate is the rate at which water enters the aquifer from the surface.

A well is a point at which water is extracted from the aquifer in a groundwater model.

A pumping rate is the rate at which water is extracted from a well in a groundwater model.

Hydraulic conductivity is a measure of the ability of a soil or rock to transmit water.

Porosity is the volume of void space in a soil or rock.

The groundwater table is the level at which the groundwater is found in the aquifer.

What is a drawdown in groundwater modelling?

Drawdown is the lowering of the groundwater table caused by pumping from a well.

What is a contaminant in groundwater modelling?

A contaminant is any substance that has entered the groundwater and may be harmful to human health or the environment.

What is a transport model in groundwater modelling?

A transport model is a model that is used to simulate the movement of contaminants in groundwater.

What is a reaction model in groundwater modelling?

A reaction model is a model that is used to simulate the chemical reactions that may occur between the contaminants and the groundwater.

What is a fate and transport model in groundwater modelling?

A fate and transport model is a model that combines transport and reaction models to simulate the movement and behavior of contaminants in groundwater.

What is a sensitivity analysis in groundwater modelling?

A sensitivity analysis is a method that is used to identify how changes in the inputs to a groundwater model affect the outputs.

What is a calibration in groundwater modelling?

Calibration is the process of adjusting the input parameters to a groundwater model to match observed data.

What is a validation in groundwater?

Validation is the process of testing a groundwater model to ensure that it accurately predicts the behavior of the groundwater system under different conditions.

What is a model uncertainty in groundwater modelling?

Model uncertainty is the degree to which a groundwater model may deviate from the true behavior of the groundwater system.

What is a parameter uncertainty in groundwater modelling?

Parameter uncertainty is the degree to which the values of the input parameters in a groundwater model are uncertain or unknown.

What is a scenario analysis in groundwater modelling?

A scenario analysis is a method that is used to test the behavior of a groundwater model under different conditions, such as changes in pumping rates or recharge rates.

What is a simulation period in groundwater modelling?

The simulation period is the period of time for which the groundwater model is run to simulate the behavior of the groundwater system.

What is a steady-state groundwater model?

A steady-state groundwater model is a model in which the groundwater system is assumed to be in a state of equilibrium, with no changes in the water level or flow over time.

What is a transient groundwater model?

A transient groundwater model is a model that takes into account changes in the groundwater system over time, such as changes in recharge rates, pumping rates, or other factors.

The model domain is the area of the aquifer that is represented in the groundwater model.

Model scale is the size of the model domain relative to the size of the actual aquifer.

Model resolution is the level of detail or granularity at which the model domain is divided into cells in a numerical groundwater model.

A model boundary is the edge of the model domain, where the boundary conditions are applied.

A model input is any parameter or data that is used as an input to a groundwater model.

A model output is the result or prediction generated by a groundwater model.

The groundwater flow equation is a mathematical equation that describes the flow of groundwater in an aquifer.

What is a mass balance equation in groundwater modelling?

A mass balance equation is a mathematical equation that describes the balance of mass (water or contaminants) in a groundwater system.

What is a calibration target in groundwater modelling?

A calibration target is the observed data that a groundwater model is calibrated against, such as water level measurements.

What is a model boundary condition in groundwater modelling?

A model boundary condition is a condition that is applied to the boundary of the model domain in a groundwater model, such as a fixed water level or a fixed flow rate.

What is a sensitivity analysis in groundwater modelling?

A sensitivity analysis is a method that is used to identify how changes in the inputs to a groundwater model affect the outputs.

What is a parameter estimation in groundwater modelling?

Parameter estimation is the process of estimating the values of the input parameters in a groundwater model based on observed data.

What is a model error in groundwater modelling?

Model error is the degree to which a groundwater model may deviate from the true behavior of the groundwater system due to simplifications or assumptions made in the model.

What is a groundwater management model?

A groundwater management model is a model that is used to optimize the management of groundwater resources, such as balancing water demand and supply or evaluating the effectiveness of management strategies.

What is a well capture zone in groundwater modelling?

A well capture zone is the area around a well where the water that is pumped comes from. It is often used in groundwater modelling to evaluate the impact of pumping on the surrounding groundwater system.

What is a contaminant transport model in groundwater modelling?

A contaminant transport model is a model that is used to simulate the movement and fate of contaminants in a groundwater system.

What is a reaction rate in groundwater modelling?

A reaction rate is the rate at which a chemical reaction occurs in a groundwater system, such as the rate at which a contaminant breaks down or reacts with other chemicals.

What is a dispersion coefficient in groundwater modelling?

A dispersion coefficient is a parameter that is used to represent the spreading of a contaminant in a groundwater system due to physical processes such as diffusion or advection.

What is a sorption coefficient in groundwater modelling?

A sorption coefficient is a parameter that is used to represent the tendency of a contaminant to be adsorbed or absorbed onto solid particles in a groundwater system.

What is a remediation strategy in groundwater modelling?

A remediation strategy is a plan or approach for cleaning up or mitigating the impact of contaminants in a groundwater system, often evaluated and designed using groundwater modelling.

Examples of groundwater models include MODFLOW, FEFLOW, SEAWAT, MT3DMS, and PEST.

MODFLOW is a numerical groundwater flow model used to simulate and predict groundwater behavior in aquifers.

FEFLOW is a finite element groundwater model that can simulate a variety of physical processes, including groundwater flow, contaminant transport, and heat transport.

SEAWAT is a numerical model that simulates the interaction between seawater and freshwater in coastal aquifers.

MT3DMS is a numerical model used for simulating the transport and fate of contaminants in groundwater systems.

PEST is a software program that uses optimization techniques to calibrate and improve the accuracy of groundwater models.

9. WATER QUALITY MODELLING

Water quality modelling is the process of creating mathematical or computational models to simulate the behavior and dynamics of water quality parameters in surface water and groundwater systems. The models use data such as water quality measurements, flow rates, and meteorological data to simulate the transport and fate of water quality parameters in a water system, and can be used to make predictions about how water quality will respond to changes in the environment or to human activities.

Water quality models can be used for a variety of purposes, such as assessing the impact of land use changes, predicting the impact of wastewater discharges, designing remediation strategies for contaminated water, and evaluating the effectiveness of water quality management plans. These models can help decision-makers to make informed choices about how to manage and protect water quality, and can provide valuable information for stakeholders such as water utilities, regulators, and conservation organizations.

Water quality modelling is a complex and interdisciplinary field, involving elements of hydrology, chemistry, and biology, as well as computer science. The models must take into account factors such as the physical and chemical properties of the water, the sources and types of contaminants that may be present, and the biological processes that affect water quality. Water quality modelling can be a powerful tool for managing and protecting water resources, but it requires careful calibration and validation to ensure that the models accurately reflect the behavior of the water system.

Water quality modelling is the process of using mathematical and computational models to predict the behavior and fate of pollutants in water systems.

The main components of a water quality model are the hydrodynamic model, the water quality model, and the data assimilation system.

A hydrodynamic model is a mathematical model that simulates the physical processes that occur in a water system, such as water flow, mixing, and turbulence.

A water quality model is a mathematical model that simulates the behavior and fate of pollutants in a water system, including transport, transformation, and fate.

A data assimilation system is a set of mathematical and computational techniques that combine observations with model simulations to improve the accuracy of model predictions.

The main types of water quality models are empirical models, conceptual models, and mechanistic models.

An empirical model is a statistical model that uses empirical relationships between variables to make predictions.

What is a conceptual model?

A conceptual model is a simplified representation of a system that captures the essential features of the system, while ignoring unnecessary details.

What is a mechanistic model?

A mechanistic model is a mathematical model that represents the physical and chemical processes that occur in a system, using fundamental principles and equations.

What are the advantages of using a mechanistic model?

The advantages of using a mechanistic model are that it can be used to predict the behavior of a system under a wide range of conditions, and it can provide insights into the underlying physical and chemical processes.

What are the limitations of using a mechanistic model?

The limitations of using a mechanistic model are that it can be complex and computationally intensive, and it may require a significant amount of data to parameterize and validate.

What are some examples of water quality models?

Some examples of water quality models are the Streeter-Phelps model, the QUAL2E model, and the WASP model.

What is the Streeter-Phelps model?

The Streeter-Phelps model is a simple mechanistic model that simulates the oxygen demand and supply in a river or stream, and predicts the oxygen concentration as a function of distance and time.

What is the QUAL2E model?

The QUAL2E model is a more complex mechanistic model that simulates the water quality of a river or stream, including temperature, dissolved oxygen, pH, nutrients, and organic matter.

What is the WASP model?

The WASP model is a comprehensive mechanistic model that simulates the water quality of a river or stream, including physical, chemical, and biological processes, and can be used to predict the effects of point and non-point source pollution.

What is the purpose of calibrating a water quality model?

The purpose of calibrating a water quality model is to adjust the model parameters so that the model predictions match the observed data as closely as possible.

What is the difference between calibration and validation?

Calibration is the process of adjusting the model parameters to match the observed data, while validation is the process of testing the model predictions against independent data.

What is the purpose of validating a water quality model?

The purpose of validating a water quality model is to test the model's ability to predict the behavior and fate of pollutants under new conditions, and to evaluate the model's accuracy and reliability.

What are some common methods for calibrating a water quality model?

Some common methods for calibrating a water quality model are manual calibration, automatic calibration using optimization algorithms, and uncertainty analysis.

What is manual calibration?

Manual calibration is the process of adjusting the model parameters manually until the model predictions match the observed data as closely as possible.

What are the advantages and disadvantages of manual calibration?

The advantages of manual calibration are that it is simple and easy to understand, and it allows the user to incorporate their expert knowledge into the model. The disadvantages are that it can be time-consuming, subjective, and may not find the optimal parameter values.

What is automatic calibration?

Automatic calibration is the process of using optimization algorithms to adjust the model parameters automatically, in order to minimize the difference between the model predictions and the observed data.

What are the advantages and disadvantages of automatic calibration?

The advantages of automatic calibration are that it can be more efficient and objective than manual calibration, and it can search for the optimal parameter values. The disadvantages are that it can be computationally intensive, and it may not always find the global optimum.

What is uncertainty analysis?

Uncertainty analysis is the process of quantifying the uncertainties in the model predictions, due to the uncertainties in the model parameters, input data, and model structure.

Why is uncertainty analysis important?

Uncertainty analysis is important because it provides information about the reliability and accuracy of the model predictions, and helps to identify the sources of uncertainty and the areas where more data or research is needed.

What are some common sources of uncertainty in water quality modelling?

Some common sources of uncertainty in water quality modelling are measurement error, parameter uncertainty, model structure uncertainty, and input data uncertainty.

What is sensitivity analysis?

Sensitivity analysis is the process of analyzing how the model predictions vary as a function of changes in the model parameters, input data, or model structure.

Why is sensitivity analysis important?

Sensitivity analysis is important because it helps to identify the most important factors that affect the model predictions, and can guide the selection of the most important model parameters or input data for further study or measurement.

What is model validation?

Model validation is the process of testing the model predictions against independent data, in order to evaluate the model's accuracy and reliability.

What are some common methods for validating a water quality model?

Some common methods for validating a water quality model are cross-validation, split-sample validation, and independent validation.

What is cross-validation?

Cross-validation is the process of dividing the data into multiple subsets, and using each subset in turn to validate the model predictions, while using the remaining data to calibrate the model.

What is split-sample validation?

Split-sample validation is the process of randomly dividing the data into two subsets, and using one subset to calibrate the model, and the other subset to validate the model predictions.

What is independent validation?

Independent validation is the process of testing the model predictions against new and independent data that were not used in the model calibration.

What are some common metrics used to evaluate the performance of a water quality model?

Some common metrics used to evaluate the performance of a water quality model are root-mean-square error, coefficient of determination, and Nash-Sutcliffe efficiency.

What is root-mean-square error?

Root-mean-square error is a measure of the difference between the model predictions and the observed data, expressed as the square root of the average of the squared differences between each predicted value and its corresponding observed value.

What is coefficient of determination?

Coefficient of determination is a statistical measure that indicates how well the model predictions fit the observed data, and is expressed as a value between 0 and 1.

What is Nash-Sutcliffe efficiency?

Nash-Sutcliffe efficiency is a statistical measure of the model performance that compares the average squared differences between the model predictions and the observed data, to the variance of the observed data.

A sensitivity coefficient is a measure of the relative change in the model predictions for a given change in one of the model parameters or input data.

A Monte Carlo simulation is a type of simulation that uses random sampling to evaluate the uncertainty and variability in a model output, based on a distribution of input values and parameter values.

Monte Carlo simulation can be used in water quality modelling to evaluate the uncertainty and variability in the model predictions, based on the uncertainty and variability in the input data and model parameters.

A decision support system is a computer-based tool that provides decision makers with information, models, and analysis tools to help them make more informed decisions.

A water quality model can be used in a decision support system to provide decision makers with information about the current state of water quality, the likely effects of different management options, and the uncertainties and risks associated with different decisions.

Some common applications of water quality modelling are in the assessment and management of water resources, the design and evaluation of wastewater treatment systems, and the evaluation of the impacts of land use change or climate change on water quality.

A 1D water quality model simulates the water quality along a one-dimensional flow path, such as a river or a pipe. A 2D water quality model simulates the water quality in a two-dimensional horizontal plane, such as a lake or a reservoir. A 3D water quality model simulates the water quality in a three-dimensional space, including vertical mixing and stratification.

The advantages of 1D water quality models are that they are simple and computationally efficient, and can be used to simulate the water quality in linear systems such as rivers and pipelines. The disadvantages are that they may not be able to capture the lateral and vertical mixing processes that occur in more complex systems.

The advantages of 2D water quality models are that they can capture the lateral and vertical mixing processes that occur in lakes and reservoirs, and can simulate the effects of wind and other surface processes. The disadvantages are that they may not be able to capture the complex flow patterns that occur in more dynamic systems.

The advantages of 3D water quality models are that they can capture the complex flow patterns, vertical mixing, and stratification that occur in more complex systems, and can simulate the effects of physical and biological processes. The disadvantages are that they are computationally intensive and require more data and calibration than simpler models.

What is a mass balance model?

A mass balance model is a type of water quality model that uses the principle of conservation of mass to simulate the fate and transport of pollutants in a water body.

How can a water quality model be used to predict the impacts of climate change on water quality?

A water quality model can be used to predict the impacts of climate change on water quality by simulating the effects of changes in temperature, precipitation, and other climate variables on water temperature, flow, and other physical and chemical processes.

What are some challenges and limitations of water quality modelling?

Some challenges and limitations of water quality modelling are the availability and quality of input data, the uncertainty and variability of model parameters, the need for calibration and validation, and the complexity and computational requirements of more advanced models. Additionally, water quality modelling may not always account for the full range of physical, chemical, and biological processes that affect water quality in the real world.

10. WATER RESOURCES MANAGEMENT

Water resources management is the process of managing and allocating water resources to meet the needs of people and the environment in a sustainable manner. It involves planning, development, distribution, and conservation of water resources to ensure that they are used efficiently and effectively.

Water resources management includes a range of activities, such as identifying and assessing water resources, developing water policies and regulations, managing water infrastructure such as dams, reservoirs, and pipelines, and monitoring and managing water quality.

Effective water resources management requires consideration of many factors, including population growth, climate change, land use patterns, and competing demands for water. It also involves balancing the needs of different stakeholders, such as agriculture, industry, cities, and the environment.

Water resources management is critical for ensuring sustainable development and protection of the environment. It requires a collaborative approach, involving stakeholders from government, industry, academia, and the public, to ensure that water resources are used in a way that is socially, economically, and environmentally sustainable.

What is water resources management?

Water resources management refers to the process of planning, developing, distributing, and managing water resources to meet human needs and protect the environment.

What are the primary sources of water?
The primary sources of water are surface water (such as rivers, lakes, and reservoirs) and groundwater (water stored underground in aquifers).

What are the main uses of water?
The main uses of water include domestic, agricultural, industrial, and environmental purposes.

What is water conservation?
Water conservation refers to the practice of using water efficiently to reduce waste and ensure sustainable use of water resources.

What is water scarcity?
Water scarcity refers to a lack of sufficient water resources to meet the needs of a particular population or area.

What is the water cycle?
The water cycle is the natural process by which water evaporates from the earth's surface, rises into the atmosphere, condenses into clouds, falls back to the earth as precipitation, and flows into rivers, lakes, and oceans.

What is a watershed?
A watershed is an area of land where all the water that falls within it drains into the same body of water, such as a river or lake.

What is a water footprint?
A water footprint is the total volume of freshwater used to produce the goods and services consumed by an individual or group.

What is water pollution?

Water pollution refers to the contamination of water resources by pollutants such as chemicals, waste products, and bacteria.

What is a water quality index?

A water quality index is a measure of the overall quality of water in a particular area based on factors such as the presence of contaminants and the health risks they pose.

What is desalination?

Desalination is the process of removing salt and other minerals from seawater or other sources of saltwater to make it suitable for human consumption or industrial use.

What is water reuse?

Water reuse refers to the practice of treating and using wastewater for non-potable purposes such as irrigation, industrial processes, and toilet flushing.

What is the role of government in water resources management?

The government plays a critical role in water resources management by setting policies and regulations, managing water infrastructure, and ensuring equitable access to water resources.

What is the role of the private sector in water resources management?

The private sector plays a role in water resources management through water supply and treatment, infrastructure development, and water management services.

What is the role of communities in water resources management?

What is the role of communities in water resources management?

Communities play a critical role in water resources management by promoting water conservation, participating in decision-making processes, and taking responsibility for their own water use.

What is the impact of climate change on water resources?

Climate change is expected to have significant impacts on water resources, including changes in precipitation patterns, increased frequency of droughts and floods, and changes in water quality.

What is water governance?

Water governance refers to the way in which water resources are managed, including the policies, institutions, and stakeholders involved in the management process.

What is integrated water resources management?

Integrated water resources management is an approach that seeks to balance competing demands for water resources while protecting the environment and ensuring sustainable use of water resources.

What is water pricing?

Water pricing refers to the use of pricing mechanisms to allocate water resources based on supply and demand and to encourage water conservation.

What is water rights?

Water rights refer to the legal rights of individuals, communities, and governments to access and use water resources.

What is water allocation?

Water allocation refers to the process of assigning water resources to various users based on their needs and priorities.

What is a water user association?

A water user association is a group of water users who work together to manage and allocate water resources in a particular area or watershed.

What is a water conservation plan?

A water conservation plan is a strategy developed by individuals, communities, or organizations to reduce water consumption and promote sustainable use of water resources.

What is water harvesting?

Water harvesting is the collection and storage of rainwater or runoff from surfaces such as roofs and paved areas for later use.

What is a water transfer?

A water transfer is the movement of water from one location to another, either through physical infrastructure such as pipelines or through agreements between water users.

What is a water balance?

A water balance is a calculation of the inputs and outputs of water in a particular area, taking into account factors such as precipitation, evapotranspiration, and runoff.

What is a water audit?

A water audit is a detailed assessment of water use and losses within a particular system or organization, often used to identify opportunities for water conservation and efficiency improvements.

What is a water efficiency standard?

A water efficiency standard is a measure of the efficiency of water use in a particular system or process, often used as a benchmark for comparison and improvement.

What is a water recycling system?

A water recycling system is a treatment process that removes impurities from wastewater and recycles it for non-potable uses such as irrigation and industrial processes.

What is a water resource assessment?

A water resource assessment is a comprehensive evaluation of the quantity and quality of water resources in a particular area, often used to inform water management decisions.

What is a groundwater management plan?

A groundwater management plan is a strategy developed by governments or water users to manage and protect groundwater resources in a particular area.

What is a water balance model?

A water balance model is a mathematical model used to simulate the movement of water within a particular system or area, often used to forecast future water availability and inform management decisions.

What is a water supply forecast?

A water supply forecast is a prediction of future water availability based on factors such as precipitation patterns, snowpack, and groundwater levels.

A drought management plan is a strategy developed by governments or water users to prepare for and respond to drought conditions, including measures to conserve water and reduce water use.

A flood management plan is a strategy developed by governments or water users to prepare for and respond to flood conditions, including measures to reduce the risk of flooding and manage floodwater.

A riparian zone is the area of land adjacent to a body of water, often characterized by a unique ecosystem and important for water quality and habitat conservation.

A water right permit is a legal authorization to use water resources for a particular purpose, typically granted by a government agency.

A water treatment plant is a facility that treats raw water to remove impurities and make it safe for human consumption.

A wastewater treatment plant is a facility that treats wastewater to remove impurities and pollutants before it is discharged back into the environment.

A stormwater management plan is a strategy developed by governments or water users to manage and reduce the impacts of stormwater runoff on water resources and the environment.

What is a water reuse permit?
A water reuse permit is a legal authorization to use treated wastewater for non-potable purposes such as irrigation and industrial processes.

What is a water conservation incentive program?
A water conservation incentive program is a program that provides financial incentives or other rewards to individuals or organizations that implement water conservation measures.

What is a water pricing policy?
A water pricing policy is a strategy developed by governments or water utilities to set prices for water usage, often designed to encourage conservation and efficient use of water resources.

What is a water footprint?
A water footprint is the amount of water used by an individual, organization, or product throughout its lifecycle, including both direct water use and the water used in the production of goods and services.

What is a virtual water trade?
Virtual water trade refers to the concept of water embodied in products that are traded internationally, taking into account the water used in their production and transport.

What is a water-use efficiency program?
A water-use efficiency program is a program that promotes the efficient use of water resources through education, outreach, and technical assistance.

What is an integrated water resources management approach?

An integrated water resources management approach is a holistic strategy that takes into account the interconnectedness of water resources and their relationship with other sectors such as agriculture, energy, and environment.

What is a water governance framework?

A water governance framework is a set of policies, laws, and institutions that govern the management and allocation of water resources in a particular area.

What is a water scarcity index?

A water scarcity index is a measurement of the availability of water resources in a particular area, taking into account factors such as population, water use, and climate.

What is a water security plan?

A water security plan is a strategy developed by governments or water users to ensure reliable access to water resources, often taking into account potential risks such as drought, climate change, and water quality issues.

11. APPLICATION OF REMOTE SENSING AND GIS

Remote sensing and GIS (Geographic Information Systems) are powerful tools for hydrologists and water resource managers. Remote sensing involves the collection of information about the Earth's surface using sensors that are mounted on aircraft or satellites, while GIS is a computer-based tool for organizing, analyzing, and displaying geographic data. In the field of hydrology and water resources, remote sensing and GIS can be used to:

- Monitor and assess changes in water resources, such as changes in river flow, lake levels, and groundwater levels, by analyzing satellite images and other remotely sensed data.

- Map and monitor land use and land cover changes that may affect water resources, such as changes in forest cover or urbanization.

- Estimate and predict water availability and demand in a given area, using GIS to analyze data on factors such as climate, topography, soil characteristics, and land use patterns.

- Identify and assess potential sources of water pollution, using remote sensing to detect changes in water quality and GIS to analyze data on potential sources of pollution.

- Plan and manage water resources infrastructure, such as dams, reservoirs, and irrigation systems, by using GIS to analyze data on factors such as topography, soil characteristics, and water availability.

Remote sensing and GIS can provide valuable information for water resource managers and decision-makers, helping them to make informed decisions about how to manage and protect water resources in a sustainable manner.

What is remote sensing?

Remote sensing is the process of acquiring information about the Earth's surface from a distance using satellite or airborne sensors.

What is GIS?

GIS (Geographic Information System) is a system designed to capture, store, manipulate, analyze, manage and present all types of spatial or geographical data.

How can remote sensing and GIS be used in hydrology?

Remote sensing and GIS can be used to monitor and manage water resources, study water quality, analyze the hydrological cycle, and identify water sources.

What types of data can be obtained through remote sensing?

Remote sensing can provide data on precipitation, soil moisture, water level, vegetation cover, land use, and more.

What is the advantage of using remote sensing in hydrology?

Remote sensing can provide a large amount of data over a wide area, which can be useful for monitoring and managing water resources.

Passive remote sensing uses sensors that measure natural radiation emitted or reflected by the Earth's surface, while active remote sensing uses sensors that emit energy and measure the response.

LIDAR (Light Detection and Ranging) is a remote sensing technique that uses laser pulses to measure distances and create detailed 3D maps of the Earth's surface.

LIDAR can be used to measure water depth, identify flood-prone areas, and create accurate topographic maps for hydrological modelling.

SAR (Synthetic Aperture Radar) is a remote sensing technique that uses microwave radiation to create images of the Earth's surface.

SAR can be used to measure soil moisture, detect changes in land surface elevation, and monitor water levels in wetlands and rivers.

NDVI (Normalized Difference Vegetation Index) is a remote sensing index that measures the amount and health of vegetation in an area.

NDVI can be used to monitor changes in vegetation cover, which can affect the hydrological cycle and water resources.

What is evapotranspiration?

Evapotranspiration is the process by which water is transferred from the Earth's surface to the atmosphere through evaporation and plant transpiration.

How can remote sensing be used to estimate evapotranspiration?

Remote sensing can provide data on land surface temperature, vegetation cover, and soil moisture, which can be used to estimate evapotranspiration.

What is a watershed?

A watershed is an area of land that drains into a common waterway, such as a river or lake.

How can GIS be used to study watersheds?

GIS can be used to map and analyze the physical and environmental characteristics of a watershed, such as land use, slope, and soil type.

What is hydrological modelling?

Hydrological modelling is the process of simulating the movement of water through the hydrological cycle using mathematical equations and computer programs.

How can GIS be used in hydrological modelling?

GIS can be used to create spatial data inputs for hydrological models, such as land cover, soil type, and topography.

What is the difference between a hydrological model and a water resources model?

A hydrological model simulates the movement of water through the hydrological cycle, while a water resources model is used to manage and allocate water resources among various uses.

How can remote sensing be used to monitor water quality?

Remote sensing can be used to monitor water quality by detecting changes in water color, temperature, and composition, as well as the presence of pollutants and algae blooms.

What is the Normalized Difference Water Index (NDWI)?

The Normalized Difference Water Index (NDWI) is a remote sensing index that measures the presence and extent of water in an area using near-infrared and shortwave infrared bands.

How can NDWI be used in hydrology?

NDWI can be used to identify the extent of surface water bodies, monitor changes in water level, and track changes in wetland areas.

What is a hydrological network?

A hydrological network is a system of interconnected rivers, lakes, and other bodies of water that together form the hydrological cycle.

How can GIS be used to model a hydrological network?

GIS can be used to create a digital elevation model (DEM) of the area and map the flow of water through the network, which can be used to simulate the movement of water through the network.

What is the difference between a digital elevation model (DEM) and a digital terrain model (DTM)?

A digital elevation model (DEM) represents the elevation of the Earth's surface, including both natural features like mountains and man-made features like buildings, while a digital terrain model (DTM) represents only the natural elevation features.

How can remote sensing be used to monitor groundwater?

Remote sensing can be used to monitor changes in the Earth's surface due to groundwater extraction, as well as the location and extent of underground aquifers.

What is the difference between passive and active groundwater monitoring?

Passive groundwater monitoring involves monitoring the natural movement and levels of groundwater, while active groundwater monitoring involves injecting tracer chemicals or electrical signals into the groundwater to monitor its movement.

How can GIS be used to manage water resources?

GIS can be used to create maps of water resources, such as surface water bodies, groundwater aquifers, and water treatment facilities, as well as to manage and allocate water resources among various uses.

What is the Water Balance Method?

The Water Balance Method is a hydrological modelling method that calculates the balance between the inputs and outputs of water in a particular area, including precipitation, evapotranspiration, runoff, and groundwater recharge.

How can GIS be used in the Water Balance Method?

GIS can be used to create a spatial database of the inputs and outputs of water in the area, which can be used to calculate the water balance using hydrological modelling software.

What is the Soil and Water Assessment Tool (SWAT)?

The Soil and Water Assessment Tool (SWAT) is a hydrological modelling tool that simulates the movement of water through the hydrological cycle, including runoff, erosion, and nutrient cycling.

How can remote sensing be used in conjunction with SWAT?

Remote sensing can provide data inputs for SWAT, such as precipitation, soil moisture, and land cover, as well as to validate and refine the model outputs.

What is the importance of water conservation?

Water conservation is important to ensure the sustainability of water resources and to reduce the negative impacts of water scarcity on the environment and human populations.

How can GIS be used to identify areas of high water use and potential water conservation measures?

GIS can be used to map areas of high-water use, such as agricultural and urban areas, and to identify potential water conservation measures, such as rainwater harvesting and drought-resistant landscaping.

What is water demand management?

Water demand management is a set of policies and practices aimed at reducing water demand and increasing water use efficiency, in order to reduce the pressure on water resources.

How can remote sensing be used to monitor water demand and use efficiency?

Remote sensing can be used to monitor changes in vegetation cover and water use patterns in agricultural and urban areas, as well as to assess the effectiveness of water conservation measures.

What is the Global Precipitation Measurement (GPM) mission?

The Global Precipitation Measurement (GPM) mission is a joint effort between NASA and the Japan Aerospace Exploration Agency (JAXA) to measure global precipitation using a constellation of satellites.

How can GPM data be used in hydrology?

GPM data can be used to monitor global precipitation patterns and to improve the accuracy of hydrological modelling and water resource management.

What is the Landsat program?

The Landsat program is a series of Earth-observing satellites that have been in operation since the 1970s, providing a long-term record of global land use and land cover changes.

How can Landsat data be used in hydrology?

Landsat data can be used to map land cover and land use changes in the context of water resources, such as changes in agricultural and urban areas and their impacts on water availability and quality.

What is the role of remote sensing in flood management?

Remote sensing can be used to monitor and predict the extent and severity of floods, as well as to map flood-prone areas and assess the impacts of floods on infrastructure and human populations.

What is the European Space Agency's (ESA) Sentinel program?

The ESA's Sentinel program is a constellation of Earth-observing satellites that provide a range of data for monitoring the environment and managing natural resources, including water resources.

Sentinel data can be used to map water resources, monitor changes in water quality and quantity, and track the impacts of climate change and human activities on water resources.

The Water Observation and Information System (WOIS) is a web-based platform developed by the European Commission's Joint Research Centre (JRC) to provide access to data and information related to water resources management.

WOIS provides a range of data and information, including satellite imagery, hydrological modelling results, and information on water scarcity and drought, that can be used to support decision-making in water resources management.

GIS can be used to map and analyze water resources data, including water quality and quantity, and to create models that simulate the movement of water through the hydrological cycle.

Challenges include the need for high-quality and accurate data, the complexity of modelling the hydrological cycle,

and the need to integrate data from multiple sources and
scales.

What is the importance of stakeholder engagement in water
resources management?

Stakeholder engagement is important for ensuring that
water resources management decisions are transparent,
equitable, and effective, and that the needs and perspectives
of all stakeholders are taken into account.

How can GIS be used to support stakeholder engagement in
water resources management?

GIS can be used to create maps and visualizations that
communicate complex water resources data to
stakeholders, as well as to create platforms for
collaborative decision-making and data sharing.

What is the future of remote sensing and GIS in hydrology
and water resources management?

The future of remote sensing and GIS in hydrology and
water resources management is likely to involve increasing
use of advanced technologies such as machine learning and
artificial intelligence, as well as increasing collaboration
between different sectors and stakeholders to address
complex water management challenges.

12. IMPACT OF CLIMATE CHANGE

Climate change has a significant impact on water resources, affecting both the quality and quantity of available water. Rising temperatures and changing precipitation patterns are causing more frequent and severe droughts, floods, and water scarcity in many regions around the world. Some of the impacts of climate change on water resources include:

Reduced snowpack and glacier melt: Changes in temperature and precipitation are causing reductions in snowpack and glacier melt, which are important sources of freshwater in many regions. This leads to decreased water availability in rivers and streams, particularly in the dry season.

Increased evapotranspiration: Higher temperatures are increasing the rate of evapotranspiration, which is the process by which water is lost from soils and plants to the atmosphere. This reduces the amount of water available for other uses, such as irrigation and drinking water.

More frequent and severe droughts: Changes in precipitation patterns are causing more frequent and severe droughts, which can lead to water shortages, reduced agricultural productivity, and economic losses.

Increased flood risk: Climate change is also causing more frequent and severe floods, particularly in regions where heavy rainfall events are becoming more common. These floods can damage infrastructure, homes, and businesses, and can lead to loss of life.

Water quality impacts: Climate change can also impact water quality, by increasing water temperature, altering flow patterns, and exacerbating pollution from agricultural and urban runoff. This can lead to harmful algal blooms, fish kills, and waterborne diseases.

The impacts of climate change on water resources are complex and vary depending on local conditions. Effective adaptation strategies are needed to ensure that water resources are managed sustainably and to mitigate the impacts of climate change on water availability and quality.

What is climate change?

Climate change is the long-term alteration of temperature and typical weather patterns in a place.

How does climate change affect water resources?

Climate change can alter the availability, quality, and quantity of water resources.

What is the impact of climate change on freshwater resources?

Climate change can cause changes in precipitation patterns, snow and ice cover, river flows, and groundwater recharge.

How does climate change affect water availability?

Climate change can lead to increased evaporation rates and reduced rainfall, leading to water scarcity in some areas.

How does climate change affect water quality?

Climate change can increase water temperatures, leading to changes in water chemistry and increases in harmful algal blooms.

How does climate change affect water storage?

Climate change can alter snow and ice melt, leading to changes in water storage in mountain regions.

What are the effects of climate change on groundwater resources?

Climate change can alter recharge rates, leading to reduced groundwater resources in some areas.

What are the effects of climate change on coastal water resources?

Climate change can lead to sea level rise, increased storm surges, and saltwater intrusion into coastal aquifers.

How does climate change affect water-related ecosystems?

Climate change can alter river flows and temperature, leading to changes in aquatic ecosystems and the species that inhabit them.

How does climate change affect water-related hazards?

Climate change can increase the frequency and intensity of floods, droughts, and other extreme weather events.

What are some of the impacts of climate change on agriculture and irrigation?

Climate change can alter precipitation patterns and soil moisture, leading to reduced crop yields and increased irrigation demands.

How does climate change affect urban water resources?

Climate change can alter water demand patterns, leading to increased water stress in some urban areas.

How does climate change affect water-related infrastructure?

Climate change can increase the risk of infrastructure damage due to floods and other extreme weather events.

How does climate change affect water-related industries, such as hydropower and fishing?

Climate change can alter river flows and water temperatures, affecting the viability of hydropower and the distribution of fish populations.

How does climate change affect water-related tourism?

Climate change can alter the availability and quality of recreational water resources, affecting the tourism industry.

What are some strategies for adapting to the impacts of climate change on water resources?

Adaptation strategies include water conservation, infrastructure improvements, and the development of drought-resistant crops.

What are some strategies for mitigating the impacts of climate change on water resources?

Mitigation strategies include reducing greenhouse gas emissions, improving water management practices, and restoring degraded watersheds.

What role do water management policies play in addressing the impacts of climate change on water resources?

Water management policies can help promote sustainable water use and protect water resources from the impacts of climate change.

What role do international agreements play in addressing the impacts of climate change on water resources?

International agreements can help promote cooperation and resource-sharing between countries facing water resource challenges.

How can communities work together to address the impacts of climate change on water resources?

Communities can work together to develop water management plans, promote water conservation, and implement infrastructure improvements.

What role do scientists play in understanding the impacts of climate change on water resources?

Scientists play a crucial role in monitoring and modelling changes in water resources and identifying strategies for adapting to and mitigating the impacts of climate change.

How can technology be used to address the impacts of climate change on water resources?

Technology can be used to improve water management practices, develop drought-resistant crops, and monitor changes in water resources.

What role do NGOs play in addressing the impacts of climate change on water resources?

NGOs can advocate for policies and practices that promote sustainable water use and protect water resources, as well as supporting local communities in adapting to and mitigating the impacts of climate change on water resources.

What are the potential economic impacts of climate change on water resources?

Climate change can lead to increased costs for water management, infrastructure repair, and drought-resistant agriculture, as well as decreased revenue from industries that rely on water resources.

How does climate change affect water availability in arid regions?

Climate change can exacerbate water scarcity in arid regions by reducing rainfall and increasing evaporation rates.

What are the potential health impacts of climate change on water resources?

Climate change can increase the risk of waterborne diseases due to changes in water quality and availability.

How does climate change affect water-related conflicts?

Climate change can increase competition for scarce water resources, leading to conflicts between different users and stakeholders.

How does climate change affect indigenous communities and their water resources?

Indigenous communities often rely heavily on traditional water sources, which may be impacted by changes in precipitation patterns, river flows, and groundwater recharge.

How does climate change affect small island developing states and their water resources?

Small island developing states are particularly vulnerable to the impacts of climate change on water resources due to their limited freshwater resources and susceptibility to sea level rise and saltwater intrusion.

What are some of the challenges of managing water resources in the face of climate change?

Challenges include predicting and adapting to uncertain changes in water availability and quality, balancing competing demands for water resources, and developing effective communication and collaboration among stakeholders.

How does climate change affect the global water cycle?

Climate change can alter the global water cycle by changing the balance of precipitation, evaporation, and runoff.

How does climate change affect glacier melt and its impact on water resources?

Climate change can accelerate glacier melt, leading to increased water flows in the short term but decreased water storage and availability in the long term.

How does climate change affect river flow patterns?

Climate change can alter river flow patterns by changing the timing and amount of precipitation and snowmelt.

How does climate change affect wetland ecosystems and their impact on water resources?

Climate change can alter the hydrology and vegetation of wetland ecosystems, impacting their ability to store and filter water.

How does climate change affect groundwater recharge rates?

Climate change can alter precipitation patterns, leading to changes in groundwater recharge rates.

How does climate change affect water temperature and its impact on aquatic ecosystems?

Climate change can increase water temperatures, leading to changes in aquatic ecosystems and the species that inhabit them.

Climate change can alter river flows and water temperatures, leading to changes in the distribution of freshwater fish populations.

Climate change can lead to water scarcity in some areas, impacting the availability of safe drinking water.

Climate change can increase the risk of harmful algal blooms and other waterborne pollutants in coastal areas.

Climate change can increase the frequency and intensity of floods and droughts, leading to more severe water-related disasters.

Climate change can alter river flows, potentially impacting the availability of water for hydropower generation.

Climate change can increase the demand for water in urban areas due to increased temperatures and the need for more frequent irrigation and cooling.

How does climate change affect the availability of water for agriculture?

Climate change can impact the availability of water for agriculture by changing precipitation patterns and reducing groundwater recharge rates.

How does climate change affect the availability of water for livestock?

Climate change can impact the availability of water for livestock by reducing water availability and quality in grazing areas.

How does climate change affect the availability of water for industrial use?

Climate change can impact the availability of water for industrial use by reducing water availability and increasing competition with other users.

How does climate change affect the availability of water for recreational use?

Climate change can impact the availability of water for recreational use by changing water levels and water quality in lakes, rivers, and oceans.

How does climate change affect the availability of water for cultural and spiritual practices?

Climate change can impact the availability of water for cultural and spiritual practices that rely on traditional water sources.

How can individuals reduce their impact on water resources in the face of climate change?

Individuals can reduce their impact on water resources by practicing water conservation, using water-efficient technologies, and supporting policies and practices that promote sustainable water use.

Governments and organizations can address the impacts of climate change on water resources by developing and implementing policies and practices that promote sustainable water use, investing in water infrastructure and management, and supporting local communities in adapting to and mitigating the impacts of climate change on water resources.

CONCLUSION

This book has covered a wide range of topics related to hydrology and water resources. We have explored the principles and practices of surface water and groundwater management, as well as the various techniques for modelling water quantity and quality.

In addition, we have discussed the application of remote sensing and GIS technology in the context of hydrology and water resources management, providing insights into how these tools can be used to enhance our understanding of the spatial and temporal dynamics of water resources.

Finally, we have also explored the impact of climate change on water resources and the challenges it presents for sustainable water management. It is clear that climate change is likely to have a significant impact on water resources globally, and particularly in India, where the monsoon plays such a critical role in the country's water balance.

Overall, this book provides a comprehensive overview of the key concepts and techniques related to hydrology and water resources. It is hoped that it will be a valuable resource for students, researchers, and practitioners working in this field, and that it will contribute to the ongoing efforts to improve the management of water resources for sustainable development.

The questions and answers provided in the book "Hydrology and Water Resources: A Comprehensive Questions and Answers Guide" cover a wide range of topics related to understanding and managing water resources. It is a useful resource for students, researchers, and professionals working in this field, as well as for individuals preparing for written tests and interviews. A chapter has been exclusively devoted for water resources of India. With its comprehensive coverage, the book is an invaluable tool for enhancing knowledge, improving performance, and gaining a competitive edge in the job market."

ABOUT THE AUTHOR

Mr. C. P. Kumar has been retired as Scientist 'G' from National Institute of Hydrology (A Government of India Society under Ministry of Jal Shakti), Roorkee - 247667 (Uttarakhand), India. His research areas of interest have been assessment of groundwater potential; seawater intrusion in coastal aquifers; numerical modelling of unsaturated flow, groundwater flow and contaminant transport; and impact of climate change on groundwater. He has authored more than 100 technical papers and reports. You may visit his website at http://www.angelfire.com/nh/cpkumar/